Harvey and the Handy Lads

Written and illustrated by Henry Brewis

FARMING PRESS

Harvey was a VERY worried hedgehog.

In less than a week, three of his cousins and two of his best friends had been squashed flat on the road by the terrible monster wagons who rumbled and roared by every day and night...

...With their big fat wheels and enormous bright eyes dazzling in the dark, they were really DANGEROUS BEASTS.

That road was definitely a noisy, smelly, scary place to be now...especially if you were a nervous little hedgehog who couldn't run very fast..!

Only yesterday Harvey himself had had a very narrow escape.

He was just halfway across the road, when a huge van missed him by no more than a whisker...the giant wheels howling past on both sides...poor little Harvey stuck in the middle...TERRIFIED!

When he'd first heard the nasty thing coming, he'd quickly rolled himself up into a ball, shut his eyes...and just hoped for the best.

...But it had been a very close shave. He knew he might not be so lucky next time.

So that evening, still shaking a little bit, Harvey went to see the brainy legal owl for some advice...

I'm sorry to trouble you,' he said, bowing to the wise old bird, 'but we hedgehogs have a really serious problem — we can hardly get over the road without being completely squashed...If it goes on like this, there'll soon be none of us left. What can we do?'

'Oh, that's easy,' said the snooty old owl. 'You shouldn't cross the road AT ALL...Indeed you're very silly even to try. If I were you I'd stay on this side for EVER, where it's perfectly safe.'

'But that's ridiculous,' exclaimed Harvey. 'We HAVE to cross sometimes ...I have relations over there...and we visit Grandma every Sunday for tea. Surely you can come up with a better solution than that!'

'Oh very well, if you MUST risk it,' said the pompous owl, with his beak in the air, 'just be sure to look both ways, wait until nothing is coming ...then run like the wind!'

Harvey was absolutely flabbergasted.

'I thought you were supposed to be brilliantly clever,' he spluttered. 'Don't you know hedgehogs simply CAN'T run fast...we've only got little short legs...and those cars and wagons can go seventy miles an hour!

'Well, cross at night when it's dark,' said the owl impatiently, 'then at least you'll see their big bright eyes coming miles away, and you can SHUFFLE over when the road is clear.'

He was gathering up his important papers now...preparing to go...'Anyway that's my expert advice,' he hooted, 'take it or leave it...and it will cost you five pounds.'

And with that, the owl flapped and fluttered away over the trees ...leaving the unhappy hedgehog with the bill.

FIVE POUNDS

'I wish I could fly,' Harvey muttered to himself. 'That would solve the problem.'

'Really...and what problem might that be I wonder?' asked a quiet voice nearby.

It was Fergus the fox, who was sneaking down to the farm in search of a tasty chicken take-away for his supper.

Now the fox was a crafty animal...maybe HE would know what to do...

So Harvey told him all about the deadly road and the monster wheely wagons. 'Five of us squashed this week already,' he sobbed. 'It's disastrous!'

'Good gracious,' said Fergus...'Why would anyone want to run over a spiky little hedgehog? – I remember once playing games with your cousin, Winston, and he just curled up into a prickly bundle, and I couldn't get anywhere near him. It was no fun at all...I had a very sore nose for days...NEVER AGAIN!'

'Well, I'm sorry about your nose,' said Harvey, 'but sadly, cousin Winston is now a dead, flattened hedgehog...and I can assure you nobody worries about a few spiky little creatures on the motorway...They just VROOM over the top of us, and we're completely squelchified..!

'So please tell me,' he asked, 'How does a fox get across that awful road?'

Fergus swished his bushy tail, 'Well it's not always easy,' he said…'but perhaps I'm a bit quicker than you, and at least so far everything has missed me…However, I DO remember coming home with a nice juicy hen one night, and very nearly running straight into a giant red tractor…WHOOSH…never saw it! Too busy dreaming about my supper I suppose.'

'Huh – you're not much help either,' sighed Harvey…'Surely SOMEBODY, SOMEWHERE in this wood has the answer..!'

'Perhaps you could JUMP over like me,' suggested a friendly frog perched on a water lily...'I might even give you a lift on my back.'

'That's very kind of you,' said Harvey, 'but no thanks — I've seen some of YOUR pals squashed on the road as well...you frogs seem to take one big jump and then sit still for a while, thinking about the next one. Knowing my luck we'd BOTH be squashed while you were sitting there thinking!'

'You could HOP over like me,' said a rabbit, as he hopped by nibbling some dandelion leaves.

'Ah well now, to be honest,' Harvey smiled, 'I've seen a few messy rabbits on the road too...Obviously you can't get across with one gigantic hop, that's for sure...And anyway, hedgehogs don't hop very well.'

'I can offer you a VERY cheap flight,' said the sneaky black crow.

But he had a peculiar glint in his eye, and Harvey was a little suspicious. The crow would probably drop him on the white line in the middle of the road...and come back later to pick up the pieces...

The bird couldn't really be trusted. Everybody knew that!

'No way,' said Harvey shaking his head…'And if I can't FLY

…or HOP…

or JUMP…

…I guess I'll just have to SHUFFLE over the road — until sooner or later one of those big bright-eyed monsters scrunches me to mincemeat..!'

So Harvey slumped down under the ash tree, feeling rather sorry for himself.

He had almost dozed off, when suddenly he heard voices. Somebody was coming through the wood, and whoever they were...THEY were obviously in a MUCH happier mood than he was. In fact they were singing merrily, and as they came nearer and nearer, Harvey could hear the words of the song quite clearly...

'Hey diddlede dee, we're busy as a bee
up with the lark, work until dark
and every job is free...

Whoops tiddleee bong, there's nothing can go wrong
As long as you smile, once in a while
and sing this silly song...'

...But the singing faded away, and stopped altogether when the happy little gang saw the sad, worried hedgehog...

Harvey looked up and recognised Ernie the badger, Nathaniel the fat brown rat, and Maxwell Mole...These were the famous Handy Lads. Everybody in the whole woodland world knew these three!

Maxwell, as always, carried the heavy bag of tools. Nat the rat struggled with a ladder over his shoulder, a bucket in one hand and a shovel in the other.

Ernie, of course, never carried ANYTHING...except his new bright-red mobile phone, a little black notebook and a pencil...because he was the boss, wasn't he? And by far the cleverest.

Indeed, Ernie was the only fellow Harvey knew who could peel an apple, and always end up with one long twisty bit of peel...no breaks at all.

And he was tremendously strong. He could even unscrew the lid from the stickiest jam jar.

And he could do really difficult sums – like seven times eleven, or fifteen divided by three...He was really BRILLIANT!

If anybody could solve the hedgehogs' problem, it would be Ernie and the Handy Lads.

...Whatever tricky job needed to be done, they were definitely the boys to call...

A garden shed for grandad rabbit's allotment for instance? – NO PROBLEM!

....A new porch for the squirrels?

...A bird bath for the sparrows?

...NOTHING was too difficult for the Handy Lads!

'You look extremely fed up,' said Ernie to the hedgehog, 'What's the matter, Harvey old friend?'

Once again Harvey told his sorry tale of death and disaster on the highway...'It's absolutely DESPERATE,' he said...'I've had some very expensive advice from the legal owl, consulted the crafty fox, had generous offers from the frog, the rabbit and the sneaky black crow...but none of them were really any help...So I STILL don't know what to do...'

Ernie climbed slowly onto a nearby tree stump to sit and give this terrible problem his full, undivided attention...

'Ummm...difficult,' he said, scratching his chin...'Let me think...'

The badger never made a really important decision in a hurry. He always thought about it very carefully...wondered about it for a while...and then considered it all again...Until, just when everybody assumed he must have forgotten the question, Ernie would come up with the answer...

So they just waited patiently.

In fact they were quite taken by surprise when Ernie suddenly leapt to his feet...

'Right,' he said, 'I believe it's really very clear...what we need is an underground tunnel!'

'A TUNNEL???' they cried in amazement.

'Exactly,' said the badger calmly...'A tunnel UNDERNEATH THE ROAD....It's obvious...Just leave it to us, Harvey...We'll begin on Monday, crack o'dawn.

And so, early on Monday morning, soon after the blackbird had woken everybody, – Harvey and the Handy Lads met behind the hedge in the meadow.

Maxwell sat on the tool bag, Nathaniel perched on his bucket...and Ernie scribbled busily in his notebook, drawing maps and plans. The big adventure was about to begin.

'The first thing we need to know,' declared Ernie, 'is precisely HOW LONG this tunnel will be...I mean it would be pointless just digging away blindly, and then coming up in the middle of the road, wouldn't it? That would be a DISASTER...So I reckon we should begin right here by this thistle, and aim to come up behind the hedge on the other side, near that sycamore tree in the cow field...Agreed?'

Everyone nodded, because nobody else had given the idea any consideration at all. They ALWAYS left the difficult details to Ernie...He was the one with the brains.

'So how far do you think it is?' asked Nathaniel.

'Well now, funny you should ask,' said Ernie...'because YOU, my dear Nat, are about to answer that very question. Firstly, however, I need to borrow your tail...'

'My T-T-TAIL?' the rat stammered anxiously.

'Oh don't worry,' Ernie smiled, 'I'm not about to chop it off...I only want to tie this piece of string to it, that's all...You'll understand in a minute. It's a very important part of the plan.'

They all watched as Ernie tied the red string to the rat's long brown tail.

'Now then, Nathaniel,' said the badger, 'I want you to scuttle through this hole in the hedge, run straight over the road...avoiding all the traffic of course...creep under the hedge on the other side, and stop right next to the sycamore tree. That will be the VERY SPOT where our tunnel will come up to the surface...Understood?'

'Understood,' said Nathaniel, and was about to dart off on his dangerous mission, when Ernie pulled him back...

'Not so fast you daft rat,' he shouted...'I'm not finished giving your orders yet. Now listen carefully...When you get there, untie the string and give it a good hard tug, — then we'll know you've made it safely...okay?'

Nathaniel nodded, saluted bravely, — and set off through the hedge, trailing the string behind him. He paused at the roadside, peering out from the long grass...looked both ways...and began his desperate dash for the far side.

Ernie, Harvey and Maxwell the mole could only wait.

Two trucks went zooming past, three cars, a motorbike and a van...and still the string was pulled through the hedge. Sometimes it stopped, then started off again as Nathaniel scurried over the road...hopefully dodging everything...

Would he survive to the other side?

Would they ever see him again?

Was the poor brave rat squashed already?

They waited and waited and waited…and then, just when even Ernie was becoming a little concerned…just when Harvey was convinced the road had claimed yet another victim…there was a tug on the string.

'That's the signal,' shouted Ernie, 'HE'S MADE IT!'

Ten long minutes later the brave rat scrambled back through the hole in the hedge, with a big grin on his ugly face.

'Mission completed,' he said breathlessly, saluting again...'So how long is this tunnel going to be, boss?'

Ernie was already measuring the string. 'From the end that was tied to your tail,' he said, 'to this knot here...and allowing a bit extra for digging down at this end, and up again at the other...

...I reckon twenty-four metres exactly.'

'WOW! - Who on earth could dig a tunnel THAT long?' asked Maxwell the mole.

'Guess...' grinned the badger.

'Oh,' said Maxwell very quietly...'Me?'

Maxwell began digging at six o'clock in the morning. They pointed him north towards the road, tied the measured length of string to one of his back legs…and down he went into the ground.

Nathaniel followed with the bucket, and every so often, he dragged it back to the surface full of soil.

The work went on all morning while Ernie did sums in his notebook and talked into his bright new phone.

Harvey made tea and jam sandwiches for the underground workers, and tried not to look worried...

Other woodland animals came to look into the hole, and play games on the growing mountain of soil.

Maxwell and Nathaniel took a short break at midday.

'How's it going?' Harvey asked anxiously when they came up into the light.

'It's not easy,' said Maxwell blinking (he didn't like the bright sunshine very much). 'We've had to go round some very big rocks, but we must be below the road now. We can hear the wagons rumbling overhead…I think we're nearly halfway.'

All through the afternoon they toiled down in the tunnel. The bucket came back again and again and again, full of soil and stones and clay...
The string attached to Maxwell's leg went further and further and further into the ground...

Harvey shuffled about nervously at the mouth of the tunnel, and every time the rat emerged he asked, 'How much longer? — Are you nearly there?'

It was almost seven o'clock in the evening when the knot on the string…the point that marked twenty-four metres…finally slid down into the hole, and Ernie knew they must have reached the sycamore tree at last.

When Nathaniel came up with his two hundredth bucketful of soil, Ernie sent him back with a message for Maxwell.

It was time to stop tunnelling, and dig UP to the surface.

The plan was, — as soon as Maxwell mole climbed out into the field on the far side, Nat the rat would hurry back along the tunnel with the good news. But as it happened, it wasn't quite as simple as that...

...When Maxwell pointed his nose skywards and tried to scratch up through the grassy roots, he came upon an enormous great lump of something blocking the way. He poked it, he punched it, but it wouldn't move!

Maxwell thought he might have heard it cough or grunt, but he wasn't sure. What could it be?

Nathaniel quickly reported the problem to Ernie…and Ernie, who wasn't going to be beaten at this stage, immediately phoned the fox.

'Ah Fergus,' he said. 'It seems Maxwell is stuck underneath a strange heavy object, and he can't get out. Could you kindly nip over the road and see what it is?'

A few anxious minutes passed before the fox rang back with the answer…
'It's a cow,' he said. 'She's just lying there right next to the tree, half asleep, chewing her cud. But don't worry…I'll soon move the dozy thing…'

And sure enough, it didn't take very long.

Fergus quietly crept up behind the dreamy old cow…and bit her sharply on the bottom…

THAT woke her up all right…and not at all sure what had happened, she leapt up and galloped away over the field to join her friends…

And so Maxwell the mole and Nat the rat were finally able to climb out of the ground into the cool evening air…and the great hedgehog tunnel was complete.

That night the forest folk held a party to celebrate.

Everybody was invited, even the sneaky black crows, and everybody came. It was the best party there'd been for years.

Hedgehogs, rabbits, frogs and mice who lived on the other side of the road, all came through the tunnel of course.

Nobody crossed the deadly road THAT night.

At midnight Ernie made a little speech, like important people often do…and with a big pair of scissors, cut the bright blue ribbon he'd draped over the entrance, specially for the occasion.

'I declare this tunnel officially OPEN!' he said in a very loud voice. And everybody clapped and cheered.

Just over the hedge, the terrible monster wagons still rumbled and roared along the road, with their big dazzling eyes and fat, spinning wheels...

...But from now on, everybody, especially all the nervous little animals who couldn't run very fast, would find a new way safely home to bed...wouldn't they?

...And Harvey was certainly the happiest hedgehog in THE WHOLE WORLD!